AF540450

Advances in Pharmacology

NIPA® GENX ELECTRONIC RESOURCES & SOLUTIONS P. LTD.
New Delhi-110 034

About the Editors

Dr. E. Sony Sharlet has obtained B.V.Sc. from NTR College of Veterinary Science, Gannavaram, Tirupati, Andhra Pradesh, India. Dr. Sony has shown more interest on Clinical Subjects namely Surgery, Gynaecology and Medicine. Dr. Sony has completed M.V.Sc. from surgery department, College of Veterinary College, Tirupati, Andhra Pradesh, India.

Dr. E. Murlinath is working as Associate Professor in Dept. of Physiology, College of Veterinary Science, Proddatur, Andhra Pradesh, India. Dr. Nah published 19 (Nineteen) research articles in peer reviewed journals. Dr. Nath published 25 (twenty five) text books. Dr. Nath received 15 Best Teaching Awards. Dr. Nath obtained 180 (One hundread eighy) review articles. Dr. Nath gained 24 (Twenty four) years of service particularly in Teaching in College of Veterinary Science at Andhra Pradesh in India. Dr. Nath published 25 Natloanl (India) books. Dr. Nath has a life membership in 6 national and international organizations. Dr. Nath published 3 Book Chapters in an International Books.

Dr. Archana Jain is working as Professor and Head, Department of Veterinary Physiology and Biochemistrey, Veterinary College, Mhow, Madhya Pradesh, India. Dr. Jain gained 32 years and 7 months of experience. Dr. Jain published so many research articles. Dr. Jain is very efficient not only in Teaching but also particularly in Research field also. Dr. Jain has command over so many branches of Physiology namely Cardio vascular Physiology, Endocrinology, Neurology, Body fluids. Excretory system and Respiratory system.

M. Guruprasad has published 200 (two hundred) articles in various journals and 25 booklets in various publications. Dr. Guruprasad is a life member of 15 national and international membership. Dr. Guruprasad has awarded with various distinguished eminent and best awards. Dr. Guruprasad also have 25 (Twenty five) PATENTS in various National and International Organizations. Dr. Guruprasad has given 195 invited talks and oral presentations to science.

Advances in Pharmacology
Volume 2

Chief Editor

E. Sony Sharlet

Executive Editor

E. Muralinath

Archana Jain

Associate Editor

M. Guruprasad

NIPA® GENX ELECTRONIC RESOURCES & SOLUTIONS P. LTD.

New Delhi-110 034

NIPA® GENX ELECTRONIC RESOURCES & SOLUTIONS P. LTD.

101,103, Vikas Surya Plaza, CU Block
L.S.C. Market, Pitam Pura, New Delhi-110 034
Ph : +91-11-43860225, Mob.: +91 9717133558, 9540816132
E-mail: newindiapublishingagency@gmail.com
Website: www.nipaersources.com

© 2025, Editors

Print ISBN: 978-93-58873-68-9
ebook ISBN: 978-93-58872-03-3

All rights reserved. No part of this publication may be reproduced, stored in a retrieval system or transmitted in any form or by any means, including electronic, mechanical, photocopying recording or otherwise without the prior written permission of the publisher or the copyright holder.

This book contains information obtained from authentic and highly reliable sources. Reasonable efforts have been made to publish reliable data and information, but the author/s, editor/s and publisher cannot assume responsibility for the validity, accuracy or completeness of all materials or information published herein or the consequences of their use. The work is published with the understanding that the publisher and author/s are not attempting to render any professional services. The author/s, editor/s and publisher have attempted to trace and acknowledge the copyright holders of all material reproduced in this publication and apologize to copyright holders if permission and/or acknowledgements to publish in this form have not been taken. If any copyrighted material has not been acknowledged, please write to us and let us know so that we may rectify the error, in subsequent reprints.

Trademark Notice: NIPA®, the NIPA® logos and their presentations (the way they are written/ presented) in this book are the trademarks of the publisher and hence may not be used without written permission, if copied or used without authorization, the infringer will be prosecuted as per law.

NIPA® also publishes books in a variety of electronic formats. Some content that appears in print may not be available in electronic books, and vice versa.

Composed and Designed by NIPA®.

Dedicated to
Mr. Nicolas Moros Moris
(President of Venezuela)

Mr. Bolsonaro
(President of Brazil)

Mr. Gabriel Boric
(President of Chile)

Mr. Daniel Ortega
(President of Nicaragua)

Sri. Narendra Modi
(Prime Minister of India)

Preface

Physiology, Pharmacology as well as Toxicology is like a fulfillment of commitment to bloom Medical, Dental and Veterinary Professors as well as students to provide a torch for Hard to crack path of competition. The main purpose of this book is to provide information regarding clinical aspects in a concise, palatable and easy to revise for competitive examinations. This book has been designed in such a way that it will not only be useful to Medical, Dental and Pharmacy students but also to the Veterinary graduates who are preparing for competitive examinations like Council of Scientific Industrial Research - National Eligibility Test (CSIR – NET), Indian Council of Agricultural Research - Junior Research Fellowship, Senior Research Fellowship and Post graduate (ICAR - JRF/SRF/PG) including other examinations conducted by Public Service Commission. This book is also useful for other competitive examinations conducted by Civil Services like IAS, IPS, IFS examinations also. It is the first attempt to provide an information regarding clinical aspects for Medical, Dental, Pharmacy and Veterinary students and we shall appeal to our learned friends and colleagues to favor us with their comments and valuable suggestions for the further improvement of this book. All authors of this book make it easier to recognize the relevance of Physiological, pharmacological and Toxicological facts for little guidance at a glance. It was a great pleasure to edit this book with the help of authors of this book also. We would like to acknowledge the suggestions of few Physiologists who helped in making this book. Some of my Physiology colleagues offered invaluable suggestions in preparing this book. Few Physiologists who helped this book progress, we express our sincere gratitude and appreciation for a job well done.

Editors

Contents

1

The Influence of Drugs on Cerebellar Abnormalities Understanding Adiadochokinesia

Abstract

The Cerebellum, a critical structure in the brain essential for coordination, balance and motor control. Can be influenced by many factors along with drugs. Adiadiochokinesia refers ro the impaired ability to perform rapid alternating movements namely rapidly pronation and supination the hands or tapping the feet. Dysfunction in Cerebellum can result in a range of motor impairment along with tremors, ataxia (lack of coordination) and dysmetria (inaccurate movements). A very few common classes of drugs associated with cerebellar dysfunction include alcohol, anti epileptic drugs, sedatives asxwell as benzodiazepines and opioid. Regarding treatment, physical therapy and rehabilitation techniques aimed at improving coordination and motor skills can be beneficial for individuals with cerebellar impairment. Finally it is concluded that drugs can exert profound effects on cerebellar function, leading to abnormalities such as adiadiochokinesia.

Introduction

The cerebellum, a critical structure in the brain essential for coordination, balance and motor control, can be influenced by various factors including drugs. One notable manifestation of cerebellar dysfunction is adiadochokinesia, a condition manifested by the inability to perform rapid alternating movements. Understanding how drugs impact cerebellar function and contribute to abnormalities like adiadochokinesia is essential for both medical professionals and the general public.

What is Adiadochokinesia?

Adiadochokinesia, also termed as dysdiadochokinesia, is a neurological sign seen in conditions impacting the cerebellum. It refers to the impaired ability to perform rapid alternating movements namely rapidly pronating and supinating the hands or tapping the feet. This inability leads to cerebellar dysfunction, disrupting the coordination and timing needed for smooth, precise movements.

The Role of the Cerebellum in Motor Function

The cerebellum plays an important role in motor coordination as well as learning. It receives sensory information from the body and integrates it with motor commands from the brain to ensure smooth, coordinated movements. Dysfunction in the cerebellum can result in a range of motor impairments, including tremors, ataxia (lack of coordination) and dysmetria (inaccurate movements).

Drugs and Cerebellar Abnormalities

Several drugs, both therapeutic and recreational, can influence cerebellar function and contribute to abnormalities like adiadochokinesia. These drugs may exert their effects in a direct manner on the cerebellum or in an indirect manner through other brain regions involved in motor control. Some common classes of drugs associated with cerebellar dysfunction include:

Alcohol

Chronic alcohol abuse is notorious for its detrimental effects specifically on the cerebellum, resulting in a condition termed as alcoholic cerebellar degeneration. This condition can manifest as ataxia, dysarthria (difficulty speaking), and adiadochokinesia, among other symptoms.

Anti Epileptic Drugs

Certain anti epileptic drugs, such as phenytoin and carbamazepine, can cause cerebellar ataxia as a side effect. These drugs may disrupt the balance of neurotransmitters in the cerebellum, impairing its function and leading to motor deficits, including adiadochokinesia.

Sedatives and Benzodiazepines

Drugs that act as central nervous system depressants, such as benzodiazepines and barbiturates, can affect cerebellar function by inhibiting neurotransmitter activity. Prolonged use of these drugs may lead to cerebellar dysfunction and associated motor impairments.

Opioids

Chronic opioid use has been linked to cerebellar abnormalities and motor dysfunction. Opioids can alter neurotransmitter levels in the brain, along with those involved in cerebellar function, leading to impaired coordination and movement control.

Treatment and Management

The management of cerebellar abnormalities like adiadochokinesia often involves addressing the underlying cause, which may include discontinuing or adjusting the dosage of drugs contributing to cerebellar dysfunction. Besides, physical therapy and rehabilitation techniques aimed at improving coordination and motor skills can be beneficial for individuals with cerebellar impairments.

Conclusion

In conclusion, drugs can exert profound effects on cerebellar function, leading to abnormalities such as adiadochokinesia. Understanding the relationship between drugs and cerebellar dysfunction is crucial for healthcare professionals to recognize and manage these conditions effectively. Furthermore, promoting awareness of the potential risks associated with drug use can help prevent cerebellar impairments and improve overall neurological health. Further research into the specific mechanisms underlying drug-induced cerebellar abnormalities is needed to develop targeted interventions and therapies for affected individuals.

References

Ackermann H, Hertrich I, Hehr T. Oral diadochokinesis in neurological dysarthrias. Folia Phoniatr Logop. 1995;47(1):15-23. [PubMed]

Algahtani HA, Fatani AN, Shirah BH, Algahtani RH. Hashimoto`s Encephalopathy Presenting with Progressive Cerebellar Ataxia. Neurosciences (Riyadh). 2019 Oct;24(4):315-319. [PMC free article] [PubMed]

Bagchi DJ, Khanna R, Raju SS. Prevalence of soft neurological signs : a study among Indian school boys. Indian J Psychiatry. 1996 Oct;38(4):196-200. [PMC free article] [PubMed]

Bodranghien F, Bastian A, Casali C, Hallett M, Louis ED, Manto M, Mariën P, Nowak DA, Schmahmann JD, Serrao M, Steiner KM, Strupp M, Tilikete C, Timmann D, van Dun K. Consensus Paper: Revisiting the Symptoms and Signs of Cerebellar Syndrome. Cerebellum. 2016 Jun;15(3):369-91. [PMC free article] [PubMed]

Brendel B, Synofzik M, Ackermann H, Lindig T, Schölderle T, Schöls L, Ziegler W. Comparing speech characteristics in spinocerebellar ataxias type 3 and type 6 with Friedreich ataxia. J Neurol. 2015 Jan;262(1):21-6. [PubMed]

Casella G, Bordo BM, Schalling R, Villanacci V, Salemme M, Di Bella C, Baldini V, Bassotti G. Neurological disorders and celiac disease. Minerva Gastroenterol Dietol. 2016 Jun;62(2):197-206. [PubMed]

Cisnéros E, Braun CM. [Vocal and respiratory diadochokinesia in Friedreich's ataxia. Neuropathological correlations]. Rev Neurol (Paris). 1995 Feb;151(2):113-23. [PubMed]

Daneault JF, Carignan B, Sadikot AF, Duval C. Inter-limb coupling during diadochokinesis in Parkinson's and Huntington's disease. Neurosci Res. 2015 Aug;97:60-8. [PubMed]

Dennis M, Salman MS, Jewell D, Hetherington R, Spiegler BJ, MacGregor DL, Drake JM, Humphreys RP, Gentili F. Upper limb motor function in young adults with spina bifida and hydrocephalus. Childs Nerv Syst. 2009 Nov;25(11):1447-53. [PMC free article] [PubMed]

Diener HC, Dichgans J. Pathophysiology of cerebellar ataxia. Mov Disord. 1992;7(2):95-109. [PubMed]

Ercoli T, Defazio G, Muroni A. Cerebellar Syndrome Associated with Thyroid Disorders. Cerebellum. 2019 Oct;18(5):932-940. [PubMed]

Ergun A, Oder W. Oral diadochokinesis and velocity of narrative speech: a prognostic parameter for the outcome of diffuse axonal injury in severe head trauma. Brain Inj. 2008 Sep;22(10):773-9. [PubMed]

Herkert PF, Hagen F, de Oliveira Salvador GL, Gomes RR, Ferreira MS, Vicente VA, Muro MD, Pinheiro RL, Meis JF, Queiroz-Telles F. Molecular characterisation and antifungal susceptibility of clinical Cryptococcus deuterogattii (AFLP6/VGII) isolates from Southern Brazil. Eur J Clin Microbiol Infect Dis. 2016 Nov;35(11):1803-1810. [PubMed]

Hermsdörfer J, Marquardt C, Wack S, Mai N. Comparative analysis of diadochokinetic movements. J Electromyogr Kinesiol. 1999 Aug;9(4):283-95. [PubMed]

Kim CK, Kalynchuk LE, Pinel JP, Kippin TE. Tolerance to the anticonvulsant and ataxic effects of pentobarbital: effect of an ascending-dose regimen. Pharmacol Biochem Behav. 1995 Dec;52(4):825-9. [PubMed]

Konstantopoulos K, Zamba-Papanicolaou E, Christodoulou K. Quantification of dysarthrophonia in a Cypriot family with autosomal recessive hereditary spastic paraplegia associated with a homozygous SPG11 mutation. Neurol Sci. 2018 Sep; 39(9): 1547-1550. [PubMed]

Montaña D, Campos-Roca Y, Pérez CJ. A Diadochokinesis-based expert system considering articulatory features of plosive consonants for early detection of Parkinson's disease. Comput Methods Programs Biomed. 2018 Feb;154:89-97. [PubMed]

Pereira AC, Brasolotto AG, Berretin-Felix G, Padovani CR. [Oral and laryngeal diadochokinesia in post cerebrovascular accident patients]. Pro Fono. 2004 Sep-Dec;16(3):283-92. [PubMed]

Pirker W, Katzenschlager R. Gait disorders in adults and the elderly : A clinical guide. Wien Klin Wochenschr. 2017 Feb;129(3-4):81-95. [PMC free article] [PubMed]

Putzhammer A, Perfahl M, Pfeiff L, Ibach B, Johann M, Zitzelsberger U, Hajak G. Performance of diadochokinetic movements in schizophrenic patients. Schizophr Res. 2005 Nov 15;79(2-3):271-80. [PubMed]

Rodriquez AA, Ford CN, Bless DM, Harmon RL. Electromyographic assessment of spasmodic dysphonia patients prior to botulinum toxin injection. Electromyogr Clin Neurophysiol. 1994 Oct-Nov;34(7):403-7. [PubMed]

Rusz J, Benova B, Ruzickova H, Novotny M, Tykalova T, Hlavnicka J, Uher T, Vaneckova M, Andelova M, Novotna K, Kadrnozkova L, Horakova D. Characteristics of motor speech phenotypes in multiple sclerosis. Mult Scler Relat Disord. 2018 Jan;19:62-69. [PubMed]

Schuelke M. Ataxia with Vitamin E Deficiency. In: Adam MP, Feldman J, Mirzaa GM, Pagon RA, Wallace SE, Bean LJH, Gripp KW, Amemiya A, editors. GeneReviews® [Internet]. University of Washington, Seattle; Seattle (WA): May 20, 2005. [PubMed]

Simoceli L, Sguillar DA, Santos HM, Caputti C. Vestibular system paresis due to emergency endovascular catheterization. Int Arch Otorhinolaryngol. 2012 Apr;16(2):282-5. [PMC free article] [PubMed]

Sinha P, Vandana VP, Lewis NV, Jayaram M, Enderby P. Evaluating the effect of risperidone on speech: A cross-sectional study. Asian J Psychiatr. 2015 Jun;15:51-5. [PubMed]

Timmermann L, Braun M, Groiss S, Wojtecki L, Ostrowski S, Krause H, Pollok B, Südmeyer M, Ploner M, Gross J, Maarouf M, Voges J, Sturm V, Schnitzler A. Differential effects of levodopa and subthalamic nucleus deep brain stimulation on bradykinesia in Parkinson's disease. Mov Disord. 2008 Jan 30;23(2):218-27. [PubMed]

Tjaden K, Watling E. Characteristics of diadochokinesis in multiple sclerosis and Parkinson's disease. Folia Phoniatr Logop. 2003 Sep-Oct;55(5):241-59. [PubMed]

Wang YT, Kent RD, Duffy JR, Thomas JE, Weismer G. Alternating motion rate as an index of speech motor disorder in traumatic brain injury. Clin Linguist Phon. 2004 Jan-Feb;18(1):57-84. [PubMed]

Wertzner HF, Pagan-Neves Lde O, Alves RR, Barrozo TF. Implications of diadochokinesia in children with speech sound disorder. Codas. 2013;25(1):52-8. [PubMed]

2

Understanding the Influence of Drugs on Cerebellar Abnormalities and Dysarthria

Sony Sharlet E.[1], Muralinath E.[2], Mohan Naidu K.[3] Srinivas Prasad [4], Jayinder Paul Singh G. [5] Pradip Kumar Das[6], Panjan Ghosh P.[7], Kinsuk Das S.[8] Kalyan C.[9], Archana Jain[10] and Guruprasad M.[11]

[1]*M.V.Sc, College of Veterinary Science, Tirupati, Andhra Pradesh, India*

[2]*College of Veterinary Science, Proddatur, Andhra Pradesh, India*

[3]*Georgia University, U.S.A.*

[4]*College of Veterinary Science, Proddatur, Amdhra Pradesh, India*

[5]*GADVASU, Ludhiana, Punjab, India*

[6]*West Bengal Univ. of Anim. & Fishery Science, Kolkata, West Bengal, India*

[7]*West Bengal Univ. of Anim. & Fishery Science, Kolkata, West Bengal, India*

[8]*West Bengal Univ. of Anim. & Fishery Science, Kolkata, West Bengal, India*

[9]*College of Veterinary Science, Andhra Pradesh, India*

[10]*Department of Veterinary Physiology and Biochemistry, Veterinary College Jabalpur, Madhya Pradesh, India*

[11]*Vaishnavi Microbial Pharma Pvt Ltd, Hyderabad, India*

Abstract

Dysarthria, a motor disorder manifested by difficulty in articulating words because of muscle weakness or incoordination, can be impacted by many factors along with drugs. Cerebellar abnormalities can lead to a variety of causes including stroke, trauma, infections and neuro degenerative diseases. Dysarthria participated refers to difficulties in controlling the muscles used for speech, leading to slurred or unintelligible speech patterns. A very few drugs acting on cerebellar function include anti epileptic drugs (AEDs), alcohol, sedatives as well as benzodiazepines and chemotherapeutic agents. In a very few cases, dose adjustments, medication substitutions or discontinuation may be necessary to alleviate symptomscand minimize further impairment also. Finally it is concluded that the influence of drugs on cerebellar function and subsequent dysarthria underscores the importance of vigilant monitoring and individualized treatment approaches.

Introduction

The cerebellum, often referred to as the "little brain," plays a crucial role in motor control, coordination, and speech. Dysarthria, a motor speech disorder manifested by difficulty in articulating words because of muscle weakness or incoordination, can be impacted by various factors along with drugs. Understanding the impact of drugs on cerebellar function and subsequent dysarthria is responsible for both medical professionals and individuals impacted by these conditions.

Cerebellar Abnormalities and Dysarthria

Cerebellar abnormalities can lead to a variety of causes including stroke, trauma, infections and neuro degenerative diseases. If the cerebellum is affected, it can lead to motor impairments namely ataxia (lack of coordination), tremors and dysarthria. Dysarthria particularly refers to difficulties in controlling the muscles used for speech, leading to slurred or unintelligible speech patterns.

Influence of Drugs on Cerebellar Function

Several drugs can influence cerebellar function either in a direct manner or an indirect manner. These drugs may impact neurotransmitter systems, interfere with neuronal activity or cause structural changes in the cerebellum. Some medications known to influence cerebellar function include:

AntiEpileptic Drugs (AEDS)

Certain antiepileptic drugs, such as phenytoin and carbamazepine, can result in cerebellar ataxia and dysarthria as side effects. These drugs may alter neuronal excitability or interfere with neurotransmitter release, influencing cerebellar function and motor coordination.

Alcohol

Chronic alcohol abuse is well-known to cause cerebellar degeneration termed as alcoholic cerebellar degeneration (ACD). Individuals with ACD may experience ataxia, tremors, and dysarthria because of the the toxic effects of alcohol on cerebellar neurons and glial cells.

Sedatives and Benzodiazepines

Drugs such as benzodiazepines (e.g., diazepam, lorazepam) and sedatives can reduce cerebellar activity, resulting in motor impairments as well as speech difficulties. Prolonged use or high doses of these medications may enhance cerebellar dysfunction and worsen dysarthria symptoms also.

Chemotherapeutic Agents

Some chemotherapy drugs, particularly those used to treat certain cancers, can exhibit neurotoxic effects on the cerebellum. Chemotherapy-induced cerebellar dysfunction may manifest as ataxia, dysarthria and other motor deficits, influencing the quality of life of cancer patients undergoing treatment.

Management and Treatment

Managing dysarthria associated with cerebellar abnormalities often involves a multidisciplinary approach, along with speech therapy, physical therapy and medication management. Speech therapy aims to enhance articulation, voice quality and overall communication skills through targeted exercises and strategies. Physical therapy focuses on addressing motor coordination and balance issues, which may improve speech intelligibility in an indirect manner.

When drugs contribute to cerebellar dysfunction and dysarthria, careful medication management is compulsory. Healthcare providers should assess the risk-benefit ratio of drug therapies, considering the potential for adverse effects particularly on cerebellar function. In a very few cases, dose adjustments, medication substitutions or discontinuation may be necessary to alleviate symptoms and minimize further impairment also.

Conclusion

The influence of drugs on cerebellar function and subsequent dysarthria underscores the importance of vigilant monitoring and individualized treatment approaches. Healthcare professionals should be aware of the potential neurotoxic effects of certain medications and consider alternative therapies when managing patients with cerebellar abnormalities. By addressing the underlying causes and optimizing treatment strategies, individuals with dysarthria can achieve improved speech outcomes and enhanced quality of life.

References

Ashizawa T, Xia G. Ataxia. Continuum (Minneap Minn). 2016 Aug;22 (4 Movement Disorders):1208-26. [PMC free article] [PubMed]

D'Arrigo S, Viganò L, Grazia Bruzzone M, Marzaroli M, Nikas I, Riva D, Pantaleoni C. Diagnostic approach to cerebellar disease in children. J Child Neurol. 2005 Nov;20(11):859-66. [PubMed]

Gudlavalleti A, Tenny S. StatPearls [Internet]. StatPearls Publishing; Treasure Island (FL): Oct 31, 2022. Cerebellar Neurological Signs. [PubMed]

Jimsheleishvili S, Dididze M. StatPearls [Internet]. StatPearls Publishing; Treasure Island (FL): Jul 24, 2023. Neuroanatomy, Cerebellum. [PMC free article] [PubMed]

Kular S, Cascella M. StatPearls [Internet]. StatPearls Publishing; Treasure Island (FL): Feb 5, 2022. Chiari I Malformation. [PubMed]

Mitoma H, Buffo A, Gelfo F, Guell X, Fucà E, Kakei S, Lee J, Manto M, Petrosini L, Shaikh AG, Schmahmann JD. Consensus Paper. Cerebellar Reserve: From Cerebellar Physiology to Cerebellar Disorders. Cerebellum. 2020 Feb;19(1):131-153. [PMC free article] [PubMed]

Patel S, Barkovich AJ. Analysis and classification of cerebellar malformations. AJNR Am J Neuroradiol. 2002 Aug;23(7):1074-87. [PMC free article] [PubMed]

Ramaekers VT, Heimann G, Reul J, Thron A, Jaeken J. Genetic disorders and cerebellar structural abnormalities in childhood. Brain. 1997 Oct;120 (Pt 10):1739-51. [PubMed]

Roostaei T, Nazeri A, Sahraian MA, Minagar A. The human cerebellum: a review of physiologic neuroanatomy. Neurol Clin. 2014 Nov;32(4):859-69. [PubMed]

Severino M, Huisman TAGM. Posterior Fossa Malformations. Neuroimaging Clin N Am. 2019 Aug;29(3):367-383. [PubMed]

Shen J, Shen J, Huang K, Wu Y, Pan J, Zhan R. Syringobulbia in Patients with Chiari Malformation Type I: A Systematic Review. Biomed Res Int. 2019;2019:4829102. [PMC free article] [PubMed]

Sun W, Li G, Lai Z, Lu Z, Lin Y, Peng J, Huang J, Hu K. Subacute Combined Degeneration of the Spinal Cord and Hydrocephalus Associated with Vitamin B12 Deficiency. World Neurosurg. 2019 Aug;128:277-283. [PubMed]

Thaller M, Hughes T. Inter-rater agreement of observable and elicitable neurological signs. Clin Med (Lond). 2014 Jun;14(3):264-7. [PMC free article] [PubMed]

Valente EM, Nuovo S, Doherty D. Genetics of cerebellar disorders. Handb Clin Neurol. 2018;154:267-286. [PubMed]

Verghese J, LeValley A, Hall CB, Katz MJ, Ambrose AF, Lipton RB. Epidemiology of gait disorders in community-residing older adults. J Am Geriatr Soc. 2006 Feb;54(2):255-61. [PMC free article] [PubMed]

3

Understanding the Influence of Drugs on Cerebellar Abnormalities Exploring the Rebound Phenomenon

Sony Sharlet E.[1], Muralinath E.[2], Mohan Naidu K.[3] Srinivas Prasad[4], Jayinder Paul Singh G[5] Pradip Kumar Das[6], Panjan Ghosh P.[7], Kinsuk Das S.[8] Kalyan C.[9], Archana Jain[10] and Guruprasad M.[11]

[1]*College of Veterinary Science, Tirupati, Andhra Pradesh, India*

[2]*College of Veterinary Science, Proddatur, Andhra Pradesh, India*

[3]*Georgia University, U.S.A.*

[4]*College of Veterinary Science, Proddatur, Amdhra Pradesh, India*

[5]*GADVASU, Ludhiana, Punjab, India*

[6]*West Bengal Univ. of Anim. & Fishery Science, Kolkata, West Bengal, India*

[7]*West Bengal Univ. of Anim. & Fishery Science, Kolkata, West Bengal, India*

[8]*West Bengal Univ. of Anim. & Fishery Science, Kolkata, West Bengal, India*

[9]*College of Veterinary Science, Andhra Pradesh, India*

[10]*Department of Veterinary Physiology and Biochemistry, Veterinary College Jabalpur, Madhya Pradesh, India*

[11]*Vaishnavi Microbial Pharma Pvt Ltd, Hyderabad, India*

Abstract

The Cerebellum plays an important role regarding coordination of movement, balance and posture. Drug _ induced Cerebellar ataxia is a general manifestation of cerebellar dysfunction. Substances namely alcohol, anti epileptic drugs and very few chemotherapy Gents can cause disruptions regarding cerebellar function, resulting in uncoordinated movements as well as impaired balance. GABAnergic and glutamatergic neuro transmission play a critical roles in cerebellar function. Upon cessation of alcohol intake, the brain undergoes compensatory changes, resulting in enhanced excitability. Withdrawl of Benzodiazepines can lead to rebound hyper excitability in the cerebellum, enhancing symptoms namely tremors as well as dysmetria. A very few anti epileptic drugs namely phenytoin and carbamazepine can lead to cerebellar symptoms. Management and Treatment include gradual Tapering, symptomatic Treatment and Rehabilitation. Finally it is concluded that the influence of drugs on cerebellar function and occurrence of rebound phenomena underscores the complexity of the brains response to pharmacological agents

Introduction

The cerebellum, a structure located at the back of the brain, plays a crucial role in coordinating movement, balance, and posture. Various factors along with drug use, can significantly impact its function resulting in cerebellar abnormalities. Among these effects, the rebound phenomenon stands out as a particularly intriguing aspect, shedding light on the intricate relationship between drugs and the brain.

Cerebellar Abnormalities and Drug Influence

Cerebellar Ataxia

Drug-induced cerebellar ataxia is a common manifestation of cerebellar dysfunction. Substances such as alcohol, anti epileptic drugs and certain chemotherapy agents can disrupt cerebellar function, leading to uncoordinated movements and impaired balance.

Neurotransmitter Imbalance

Many drugs alter neurotransmitter levels in the brain, affecting communication between neurons. GABAergic and glutamatergic neuro transmission, in particular, play crucial roles in cerebellar function. Disruption of these systems by drugs can contribute to cerebellar abnormalities.

Drug Withdrawal

The rebound phenomenon occurs when the brain attempts to restore equilibrium following the cessation of drug use. In the context of cerebellar function, this rebound effect can manifest as exaggerated symptoms of cerebellar dysfunction namely tremors and ataxia, during withdrawal from certain substances.

Understanding the Rebound Phenomenon

Alcohol

Chronic alcohol consumption suppresses excitatory neuro transmission in the cerebellum. Upon cessation of alcohol intake, the brain undergoes compensatory changes, leading to increased excitability. This rebound hyper excitability contributes to symptoms of alcohol withdrawal, including tremors and ataxia.

Benzodiazepines

These drugs enhance the inhibitory effects of GABA, leading to sedation and muscle relaxation. Withdrawal from benzodiazepines can result in rebound hyperexcitability in the cerebellum, exacerbating symptoms such as tremors and dysmetria.

Anti Epileptic Drugs

Some anti epileptic drugs, such as phenytoin and carbamazepine, can cause cerebellar dysfunction. Abrupt withdrawal from these medications can trigger rebound seizures and exacerbate cerebellar symptoms.

Management and Treatment

Gradual Tapering

To mitigate the risk of rebound phenomena, clinicians often recommend gradual tapering of medications rather than abrupt cessation, allowing the brain to adjust gradually to changes in neurotransmitter levels.

Symptomatic Treatment

Patients experiencing rebound symptoms may require symptomatic treatment to alleviate discomfort and manage complications. This may include pharmacological interventions to control tremors and ataxia, as well as supportive measures to address psychological symptoms.

Rehabilitation

For individuals with persistent cerebellar abnormalities, rehabilitation programs focused on improving coordination, balance, and motor skills can be beneficial. Physical therapy, occupational therapy, and speech therapy may be employed to address specific deficits.

Conclusion

The influence of drugs on cerebellar function and the occurrence of rebound phenomena underscore the complexity of the brain's response to pharmacological agents. By understanding these mechanisms, clinicians can better anticipate and manage cerebellar abnormalities associated with drug use, ultimately improving patient outcomes and quality of life. Continued research into the neurobiology of addiction and withdrawal will further enhance our understanding of these phenomena and inform the development of more effective therapeutic strategies.

References

Baker KG, Harding AJ, Halliday GM, Kril JJ, Harper CG. Neuronal loss in functional zones of the cerebellum of chronic alcoholics with and without Wernicke's encephalopathy. Neuroscience. 1999;91(2):429-38. [PubMed]

Bötzel K, Tronnier V, Gasser T. The differential diagnosis and treatment of tremor. Dtsch Arztebl Int. 2014 Mar 28;111(13):225-35; quiz 236. [PMC free article] [PubMed]

Deuschl G, Wenzelburger R, Löffler K, Raethjen J, Stolze H. Essential tremor and cerebellar dysfunction clinical and kinematic analysis of intention tremor. Brain. 2000 Aug;123 (Pt 8):1568-80. [PubMed]

Kestenbaum M, Michalec M, Yu Q, Pullman SL, Louis ED. Intention Tremor of the Legs in Essential Tremor: Prevalence and Clinical Correlates. Mov Disord Clin Pract. 2015 Mar 01;2(1):24-28. [PMC free article] [PubMed]

Leegwater-Kim J, Louis ED, Pullman SL, Floyd AG, Borden S, Moskowitz CB, Honig LS. Intention tremor of the head in patients with essential tremor. Mov Disord. 2006 Nov;21(11):2001-5. [PubMed]

Lenka A, Louis ED. Revisiting the Clinical Phenomenology of "Cerebellar Tremor": Beyond the Intention Tremor. Cerebellum. 2019 Jun;18(3):565-574. [PubMed]

Louis ED, Frucht SJ, Rios E. Intention tremor in essential tremor: Prevalence and association with disease duration. Mov Disord. 2009 Mar 15;24(4):626-7. [PMC free article] [PubMed]

Louis ED. Tremor. Continuum (Minneap Minn). 2019 Aug;25(4):959-975. [PubMed]

McCreary JK, Rogers JA, Forwell SJ. Upper Limb Intention Tremor in Multiple Sclerosis: An Evidence-Based Review of Assessment and Treatment. Int J MS Care. 2018 Sep-Oct; 20(5):211-223. [PMC free article] [PubMed]

Poser CM, Brinar VV. Diagnostic criteria for multiple sclerosis. Clin Neurol Neurosurg. 2001 Apr;103(1):1-11. [PubMed]

Raju SS, Niranjan A, Monaco EA, Flickinger JC, Lunsford LD. Stereotactic radiosurgery for medically refractory multiple sclerosis-related tremor. J Neurosurg. 2018 Apr; 128(4):1214-1221. [PubMed]

Sternberg EJ, Alcalay RN, Levy OA, Louis ED. Postural and Intention Tremors: A Detailed Clinical Study of Essential Tremor vs. Parkinson's Disease. Front Neurol. 2013;4:51. [PMC free article] [PubMed]

Torvik A, Torp S. The prevalence of alcoholic cerebellar atrophy. A morphometric and histological study of an autopsy material. J Neurol Sci. 1986 Aug;75(1):43-51. [PubMed]

Wishart HA, Roberts DW, Roth RM, McDonald BC, Coffey DJ, Mamourian AC, Hartley C, Flashman LA, Fadul CE, Saykin AJ. Chronic deep brain stimulation for the treatment of tremor in multiple sclerosis: review and case reports. J Neurol Neurosurg Psychiatry. 2003 Oct;74(10):1392-7. [PMC free article] [PubMed]

Zakaria R, Lenz FA, Hua S, Avin BH, Liu CC, Mari Z. Thalamic physiology of intentional essential tremor is more like cerebellar tremor than postural essential tremor. Brain Res. 2013 Sep 05;1529:188-99. [PMC free article] [PubMed]

4

Understanding Hydrocephalus Causes Symptoms, Diagnosis Differential Diagnosis and Treatment

Sony Sharlet E.[1]***, Muralinath E.***[2]***, Sai Hemachand N.***[3]
Sravani K [4] ***and Guru Prasad M.***[5]

[1]*College of Veterinary Science, Tirupati, Andhra Pradesh, India*
[2]*College of Veterinary Science, Proddatur, Andhra Pradesh, India*
[3]*Internee, Veterinary Poly Clinic, Nellore, Andhra Pradesh, India*
[4]*Nuzeveedu, Andhra Pradesh, India*
[5]*Vaishnavi Microbial Phama Pvt. Ltd., Hyderabad, India*

Abstract

An accumulation of cerebro Spinal fluid (CSF) within the brains ventricles, leading to the enhancement of intra cranial pressure (ICP). This condition can happen at any age and may result from many underlying causes. Hydrocephalus can be congenital or acquired. Symptoms of Hydrocephalus include headache, nausea, vomiting, Gait disturbances, enlargement of the head in infants. Diagnosis is based on medical history, review, physical examination, Diagnostic tests and imaging studies namely MRI or CT scans. Differential Diagnosis include normal pressure Hydrocephalus, intea cranial hemorrhage and idiopathic intra cranial hypertension. Treatment is dependent upon ventricular shunt and endoscopic third ventriculostomy. Finally it is concluded that Hydrocephalus is a complex neurological condition along with diverse causes as well as presentations.

Introduction

Hydrocephalus, often referred to as "water on the brain," is a condition manifested by the accumulation of cerebrospinal fluid (CSF) within the brain's ventricles, resulting in the enhancement of intracranial pressure (ICP). This condition can happen at any age and may result from various underlying causes. Understanding its history, causes, symptoms, diagnosis, differential diagnosis and treatment is critical for effective management.

History

Hydrocephalus has been recognized for centuries, with historical references dating back to ancient times. The term "hydrocephalus" itself comes from the Greek words "hydro" (water) and "cephalus" (head). Throughout history, various theories have emerged regarding its causes and treatments. Early treatments often involved surgical interventions aimed at draining excess fluid from the brain. Over time, advancements in medical imaging, neurosurgery techniques, and understanding of the underlying patho physiology have shown good results regarding the improvement oft he management of hydrocephalus.

Causes

Hydrocephalus can be congenital (present at birth) or acquired (developing later in life). Common causes include:

1. **Congenital:** Abnormal development of the brain's ventricular system during fetal growth, genetic factors, prenatal infections, maternal drug or alcohol abuse.
2. **Acquired:** Infections (such as meningitis or brain abscess), tumors, obstructing CSF flow, traumatic brain injury, intra ventricular hemorrhage (often seen in premature infants) and complications of certain medical conditions like Chiari malformation or Dandy-Walker syndrome.

Symptoms

The symptoms of hydrocephalus can vary depending on factors such as age, rate of fluid accumulation and underlying cause. Common symptoms include: Headaches, Nausea and vomiting, Blurred or double vision, Changes in mental status or behavior, Difficulty walking or balancing (gait disturbances) and Enlargement of the head in infants (due to open cranial sutures)

Diagnosis

Diagnosing hydrocephalus typically involves a combination of medical history, review, physical examination and diagnostic tests. These may include: Neurological examination to assess reflexes, muscle strength and coordination.

Imaging studies such as MRI (Magnetic Resonance Imaging) or CT (Computed Tomography) scans to visualize the brain's ventricular system and identify any abnormalities.

Lumbar puncture (spinal tap) to analyze the composition and pressure of CSF.

Differential Diagnosis

Several conditions may present with symptoms similar to hydrocephalus, making it essential to consider other possibilities. Differential diagnoses may include:

Normal Pressure Hydrocephalus (NPH)

Manifested by by the classic triad of gait disturbances, cognitive impairment and urinary incontinence in elderly individuals.

Brain tumor: Mass effect from tumors can lead to increased ICP and neurological symptoms.

Intra Cranial Hemorrhage

Bleeding within the brain can cause similar symptoms to hydrocephalus, especially in traumatic cases.

Idiopathic Intra Cranial Hypertension (*Pseudomotor cerebri*)

Elevated ICP without apparent cause, frequently seen in overweight women of childbearing age.

Treatment

Treatment for hydrocephalus aims to reduce intracranial pressure and manage symptoms. Options may include:

Ventricular Shunt

A surgical procedure where a thin tube (shunt) is implanted to divert excess CSF from the brain's ventricles to another part of the body such as the abdominal cavity.

Endoscopic Third Ventriculostomy (ETV)

A minimally invasive procedure where a neuro endoscope is used to create a new pathway for CSF to bypass obstruction and flow out of the brain's ventricles.

Medications

In some cases, medications may be prescribed to reduce CSF production or reduce brain swelling. Regular monitoring and follow-up to adjust treatment as needed and manage potential complications.

Conclusion

In conclusion, hydrocephalus is a complex neurological condition along with diverse causes and presentations. Early diagnosis and appropriate management play an important role to prevent complications and improve outcomes for affected individuals. Advances in medical technology and research continue to enhance our understanding and treatment options for this condition, offering hope for those living with hydrocephalus.

References

Bakkour A, Morris JC, Wolk DA, Dickerson BC. The effects of aging and Alzheimer's disease on cerebral cortical anatomy: specificity and differential relationships with cognition. Neuroimage. 2013 Aug 01;76:332-44. [PMC free article] [PubMed]

Bidot S, Saindane AM, Peragallo JH, Bruce BB, Newman NJ, Biousse V. Brain Imaging in Idiopathic Intracranial Hypertension. J Neuroophthalmol. 2015 Dec;35(4):400-11. [PubMed]

Chen S, Luo J, Reis C, Manaenko A, Zhang J. Hydrocephalus after Subarachnoid Hemorrhage: Pathophysiology, Diagnosis, and Treatment. Biomed Res Int. 2017;2017:8584753. [PMC free article] [PubMed]

Damkier HH, Brown PD, Praetorius J. Cerebrospinal fluid secretion by the choroid plexus. Physiol Rev. 2013 Oct;93(4):1847-92. [PubMed]

Eymann R. [Clinical symptoms of hydrocephalus]. Radiologe. 2012 Sep;52(9):807-12. [PubMed]

Fink KR, Benjert JL. Imaging of Nontraumatic Neuroradiology Emergencies. Radiol Clin North Am. 2015 Jul;53(4):871-90, x. [PubMed]

Garne E, Loane M, Addor MC, Boyd PA, Barisic I, Dolk H. Congenital hydrocephalus--prevalence, prenatal diagnosis and outcome of pregnancy in four European regions. Eur J Paediatr Neurol. 2010 Mar;14(2):150-5. [PubMed]

Hamilton MG. Treatment of hydrocephalus in adults. Semin Pediatr Neurol. 2009 Mar;16(1):34-41. [PubMed]

Isaacs AM, Riva-Cambrin J, Yavin D, Hockley A, Pringsheim TM, Jette N, Lethebe BC, Lowerison M, Dronyk J, Hamilton MG. Age-specific global epidemiology of hydrocephalus: Systematic review, metanalysis and global birth surveillance. PLoS One. 2018;13(10):e0204926. [PMC free article] [PubMed]

Kahlon B, Annertz M, Ståhlberg F, Rehncrona S. Is aqueductal stroke volume, measured with cine phase-contrast magnetic resonance imaging scans useful in predicting outcome of shunt surgery in suspected normal pressure hydrocephalus? Neurosurgery. 2007 Jan;60(1):124-9; discussion 129-30. [PubMed]

Kartal MG, Algin O. Evaluation of hydrocephalus and other cerebrospinal fluid disorders with MRI: An update. Insights Imaging. 2014 Aug;5(4):531-41. [PMC free article] [PubMed]

Kim H, Jeong EJ, Park DH, Czosnyka Z, Yoon BC, Kim K, Czosnyka M, Kim DJ. Finite element analysis of periventricular lucency in hydrocephalus: extravasation or transependymal CSF absorption? J Neurosurg. 2016 Feb;124(2):334-41. [PubMed]

Langner S, Fleck S, Baldauf J, Mensel B, Kühn JP, Kirsch M. Diagnosis and Differential Diagnosis of Hydrocephalus in Adults. Rofo. 2017 Aug;189(8):728-739. [PubMed]

Larsson A, Moonen M, Bergh AC, Lindberg S, Wikkelsö C. Predictive value of quantitative cisternography in normal pressure hydrocephalus. Acta Neurol Scand. 1990 Apr;81(4):327-32. [PubMed]

LeMay M, Hochberg FH. Ventricular differences between hydrostatic hydrocephalus and hydrocephalus ex vacuo by computed tomography. Neuroradiology. 1979 Apr 26;17(4):191-5. [PubMed]

Markey KA, Mollan SP, Jensen RH, Sinclair AJ. Understanding idiopathic intracranial hypertension: mechanisms, management, and future directions. Lancet Neurol. 2016 Jan;15(1):78-91. [PubMed]

Munch TN, Rostgaard K, Rasmussen ML, Wohlfahrt J, Juhler M, Melbye M. Familial aggregation of congenital hydrocephalus in a nationwide cohort. Brain. 2012 Aug;135(Pt 8):2409-15. [PubMed]

Pini L, Pievani M, Bocchetta M, Altomare D, Bosco P, Cavedo E, Galluzzi S, Marizzoni M, Frisoni GB. Brain atrophy in Alzheimer's Disease and aging. Ageing Res Rev. 2016 Sep;30:25-48. [PubMed]

Preuss M, Hoffmann KT, Reiss-Zimmermann M, Hirsch W, Merkenschlager A, Meixensberger J, Dengl M. Updated physiology and pathophysiology of CSF circulation--the pulsatile vector theory. Childs Nerv Syst. 2013 Oct;29(10):1811-25. [PubMed]

Ragan DK, Cerqua J, Nash T, McKinstry RC, Shimony JS, Jones BV, Mangano FT, Holland SK, Yuan W, Limbrick DD. The accuracy of linear indices of ventricular volume in pediatric hydrocephalus: technical note. J Neurosurg Pediatr. 2015 Jun;15(6):547-51. [PMC free article] [PubMed]

Rekate HL. A contemporary definition and classification of hydrocephalus. Semin Pediatr Neurol. 2009 Mar;16(1):9-15. [PubMed]

Richards JE, Sanchez C, Phillips-Meek M, Xie W. A database of age-appropriate average MRI templates. Neuroimage. 2016 Jan 01;124(Pt B):1254-1259. [PMC free article] [PubMed]

Rojas R, Riascos R, Vargas D, Cuellar H, Borne J. Neuroimaging in drug and substance abuse part I: cocaine, cannabis, and ecstasy. Top Magn Reson Imaging. 2005 Jun;16(3):231-8. [PubMed]

Tully HM, Dobyns WB. Infantile hydrocephalus: a review of epidemiology, classification and causes. Eur J Med Genet. 2014 Aug;57(8):359-68. [PMC free article] [PubMed]

Vinchon M, Rekate H, Kulkarni AV. Pediatric hydrocephalus outcomes: a review. Fluids Barriers CNS. 2012 Aug 27;9(1):18. [PMC free article] [PubMed]

5

Understanding Drugs Acting on Cerebellar Disorders Targeting Intention Tremor

Sony Sharlet E.[1], Muralinath E.[2], Sai Hemachand N.[3] Sravani K.[4] and Guru Prasad M.[5]

[1]College of Veterinary Science, Tirupati, Andhra Pradesh, India

[2]College of Veterinary Science, Proddatur, Andhra Pradesh, India

[3]Internee, Veterinary Poly Clinic, Nellore, Andhra Pradesh, India

[4]Nuzeveedu, Andhra Pradesh, India

[5]Vaishnavi Microbial Phama Pvt. Ltd., Hyderabad, India

Abstract

The cerebellum plays an important role regard6 motor control, coordination and balance. If cerebellum experience dysfunction, it can result in various movement disorders, one of which is inhibition tremors. In the Cerebellum, GABAnergic neurons play an important role in modulating motor activities as well as coordination. By enhancing inhibitory neuro transmission, Benzo diazepines can assist on alleviating inhibition tremor symptoms. Baclofen is used to manage spasticity and can also provide relief especially for intention tremor. By blocking beta_ adrenergic receptors, propranolol reduces sympathetic nervous system activity, there. By reducing tremor activity

Pyrimidine, an anticonvulsant enhances GABAnergic neuro transmission. Targeted manipulation of gene expression within the Cerebellum mau offer more precise ND sustainable approach to manage inhibition tremor. Finally it is concluded that inhibition tremor, stemming from cerebellar disorder, exhibits significant challenges to individuals motor function as well as quality of life.

Introduction

The cerebellum, a critical structure located at the base of the brain, plays an important role in motor control, coordination and balance. When the cerebellum experiences dysfunction, it can lead to various movement disorders, one of which is inhibition tremor. Inhibition tremor is exhibited as an involuntary shaking or trembling of body particularly during voluntary movements, impairing fine motor skills and coordination. While there is no cure for cerebellar disorders, pharmacotherapy offers options to manage symptoms, including inhibition tremor.

The Role of Neurotransmitter in Cerebellar Dysfunction

To understand how drugs target inhibition tremor, it's essential to grasp the neurotransmitter systems involved. GABA (gamma-aminobutyric acid) is the primary inhibitory neurotransmitter in the brain, exerting its effects by binding to GABA receptors. In the cerebellum, GABAergic neurons play a crucial role in modulating motor functions and coordination. Dysfunction in these pathways can lead to inhibition tremor.

Pharmaco Therapy for Inhibition Tremor

GABAnergic Drugs

Benzo Diazepines

Medications like clonazepam enhance the effects of GABA by binding to specific GABA-A receptor sites. By increasing inhibitory neurotransmission, benzodiazepines can help alleviate inhibition tremor symptoms.

Baclofen

This GABA-B receptor agonist behaves by reducing excitatory neurotransmitter release, thus dampening abnormal neuronal activity in the cerebellum. Baclofen is often used to manage spasticity and can also provide relief particularly for inhibition tremor.

Beta Blockers

Propranolol

While primarily used to treat cardiovascular conditions, propranolol has found utility in managing tremors, including those linked to cerebellar dysfunction. By blocking beta-adrenergic receptors, propranolol decreases sympathetic nervous system activity, thereby diminishing tremor severity.

Anti Convulsants

Primidone

Originally developed as an anticonvulsant, primidone has exhibited efficacy in managing various tremor disorders, including inhibition tremor. Its mechanism of action is related to an enhancement of GABAnergic neurotransmission and possibly sodium channel blockade.

Gabapentin and Pregabalin

These medications, primarily indicated for neuropathic pain and epilepsy, also exhibit modulatory effects on GABAnergic transmission. By increasing GABA levels in the brain, they can attenuate inhibition tremor symptoms.

Challenges and Consideration

While pharmacotherapy can provide symptomatic relief for inhibition tremor, several challenges exist:

Side Effects

Many drugs used to manage inhibition tremor can cause sedation, dizziness, and cognitive impairment, limiting their tolerability and long-term use.

Individual Variability

Response to medication varies among individuals, necessitating a tailored approach to treatment.

Limited Efficacy

Some patients may experience incomplete symptom control or develop tolerance to medication over time, prompting the exploration of alternative therapies.

Future Directions

As our understanding of cerebellar disorders deepens, researchers are exploring novel therapeutic avenues, including:

Neuro Modulation Techniques

Deep brain stimulation and trans cranial magnetic stimulation hold promise for modulating abnormal neuronal activity in the cerebellum.

Gene Therapy

Targeted manipulation of gene expression within the cerebellum may offer a more precise and sustainable approach to managing inhibition tremor.

Emerging Pharmacotherapies

Ongoing research aims to identify compounds that selectively target cerebellar dysfunction while minimizing off-target effects.

Conclusion

Inhibition tremor, stemming from cerebellar dysfunction, shows significant challenges to individuals' motor function and quality of life. While pharmacotherapy remains a cornerstone of management, its efficacy is often accompanied by side effects and variability in response. Nevertheless, ongoing research efforts offer hope for more targeted and effective treatments, ultimately improving outcomes for those affected by cerebellar disorders and inhibition tremor.

References

Baker KG, Harding AJ, Halliday GM, Kril JJ, Harper CG. Neuronal loss in functional zones of the cerebellum of chronic alcoholics with and without Wernicke's encephalopathy. Neuroscience. 1999;91(2):429-38. [PubMed]

Bötzel K, Tronnier V, Gasser T. The differential diagnosis and treatment of tremor. Dtsch Arztebl Int. 2014 Mar 28;111(13):225-35; quiz 236. [PMC free article] [PubMed]

Deuschl G, Wenzelburger R, Löffler K, Raethjen J, Stolze H. Essential tremor and cerebellar dysfunction clinical and kinematic analysis of intention tremor. Brain. 2000 Aug;123 (Pt 8):1568-80. [PubMed]

Kestenbaum M, Michalec M, Yu Q, Pullman SL, Louis ED. Intention Tremor of the Legs in Essential Tremor: Prevalence and Clinical Correlates. Mov Disord Clin Pract. 2015 Mar 01;2(1):24-28. [PMC free article] [PubMed]

Leegwater-Kim J, Louis ED, Pullman SL, Floyd AG, Borden S, Moskowitz CB, Honig LS. Intention tremor of the head in patients with essential tremor. Mov Disord. 2006 Nov;21(11):2001-5. [PubMed]

Lenka A, Louis ED. Revisiting the Clinical Phenomenology of "Cerebellar Tremor": Beyond the Intention Tremor. Cerebellum. 2019 Jun;18(3):565-574. [PubMed]

Louis ED, Frucht SJ, Rios E. Intention tremor in essential tremor: Prevalence and association with disease duration. Mov Disord. 2009 Mar 15;24(4):626-7. [PMC free article] [PubMed]

Louis ED. Tremor. Continuum (Minneap Minn). 2019 Aug;25(4):959-975. [PubMed]

McCreary JK, Rogers JA, Forwell SJ. Upper Limb Intention Tremor in Multiple Sclerosis: An Evidence-Based Review of Assessment and Treatment. Int J MS Care. 2018 Sep-Oct;20(5):211-223. [PMC free article] [PubMed]

Poser CM, Brinar VV. Diagnostic criteria for multiple sclerosis. Clin Neurol Neurosurg. 2001 Apr;103(1):1-11. [PubMed]

Raju SS, Niranjan A, Monaco EA, Flickinger JC, Lunsford LD. Stereotactic radiosurgery for medically refractory multiple sclerosis-related tremor. J Neurosurg. 2018 Apr;128(4):1214-1221. [PubMed]

Sternberg EJ, Alcalay RN, Levy OA, Louis ED. Postural and Intention Tremors: A Detailed Clinical Study of Essential Tremor vs. Parkinson's Disease. Front Neurol. 2013;4:51. [PMC free article] [PubMed]

Torvik A, Torp S. The prevalence of alcoholic cerebellar atrophy. A morphometric and histological study of an autopsy material. J Neurol Sci. 1986 Aug;75(1):43-51. [PubMed]

Wishart HA, Roberts DW, Roth RM, McDonald BC, Coffey DJ, Mamourian AC, Hartley C, Flashman LA, Fadul CE, Saykin AJ. Chronic deep brain stimulation for the treatment of tremor in multiple sclerosis: review and case reports. J Neurol Neurosurg Psychiatry. 2003 Oct;74(10):1392-7. [PMC free article] [PubMed]

Zakaria R, Lenz FA, Hua S, Avin BH, Liu CC, Mari Z. Thalamic physiology of intentional essential tremor is more like cerebellar tremor than postural essential tremor. Brain Res. 2013 Sep 05;1529:188-99. [PMC free article] [PubMed]

6

Understanding Cerebellar Disorders and Asynergia An Overview of Drug Therapie

Sony Sharlet E.[1], Muralinath E.[2], Sai Hemachand N.[3] Sravani K.[4] and Guru Prasad M.[5]

[1]College of Veterinary Science, Tirupati, Andhra Pradesh, India

[2]College of Veterinary Science, Proddatur, Andhra Pradesh, India

[3]Internee, Veterinary Poly Clinic, Nellore, Andhra Pradesh, India

[4]Nuzeveedu, Andhra Pradesh, India

[5]Vaishnavi Microbial Phama Pvt. Ltd., Hyderabad, India

Abstract

The Cerebellum plays an important role especially in coordinating voluntary movements as well as controlling balance. If Cerebellum is affected by abnormalities namely asynergia, these essential functions can be impaired, resulting in a significant difficulties regarding movement and coordination. Asynergia is related to lack of ability regarding coordination of muscle movements in a smooth manner, leading to Jerry or disjointed motions. The medications can modulate neuro transmitter activity, decrease symptoms namely tremors and rigidity and increase motor activity. Levpdopa is generally prescribed to decrease symptoms cerebellar abnormalities. Dopamine plays an important role in controlling movement and it's deficiency is implicated particularly in conditions namely Parkinsons disease, which can present with cerebellar dysfunction. Levodopa, frequently combined with carbi dopa to increase it's effectiveness, can help regarding an enhancement of motor activity and decrease tremors in patients with cerebellar disorders along with asynergia. Drugs that increase GABAnergic transmission, namely benzodiazepines and baclofen are generally used to manage symptoms namely spasticity associated with cerebellar abnormalities. Baclofen, a GABA _ receptor agonist, can decrease muscle spasticity and enhance coordination particularly in patients with cerebellar abnormalities. Anti cholinergic drugs namely trihexyphenidyl, can assist regarding reduction of symptoms namely tremors and rigidity by stopping the effects of acetyl choline. Dopamine receptor agonist namely pramipexole and ropinirole are useful regarding management of motor symptoms particularly in cerebellar abnormalities. Muscle relaxants namely tizanidine and dantrolene are used to decrease muscle stiffness and spasticity especially on patients with cerebellar abnormality. Finally it us concluded that pharmaco therapy can reduce symptoms and enhance quality of life for individuals with cerebellar abnormalities namely asynergia.

Introduction

The cerebellum, located at the base of the brain, plays a crucial role in coordinating voluntary movements, maintaining posture, and regulating balance. When the cerebellum is affected by disorders, such as asynergia, these essential functions can be impaired, resulting in significant difficulties in movement and coordination. Asynergia, specifically, is related to the lack of ability regarding coordination of muscle movements smoothly, resulting in jerky or disjointed motions.

While there is no single pharmacological cure for cerebellar disorders like asynergia, several drugs can help manage symptoms and improve patients' quality of life. These medications aim to modulate neurotransmitter activity, reduce symptoms such as tremors and rigidity, and enhance motor function. Let's delve into some of the key drug therapies used in the management of asynergia and other cerebellar disorders.

Dopaminergic Agents

Drugs that enhance dopamine activity in the brain, such as levodopa, are commonly prescribed to decrease symptoms of cerebellar disorders. Dopamine plays an important role in regulating movement, and its deficiency is implicated in conditions like Parkinson's disease, which can present with cerebellar dysfunction.

Levodopa, often combined with carbidopa to enhance its effectiveness, can help improve motor function and reduce tremors in patients with cerebellar disorders, including asynergia.

GABAnergic Agents

Gamma-aminobutyric acid (GABA) is the primary inhibitory neurotransmitter in the central nervous system and plays a critical l role in modulating neuronal activity. Drugs that enhance GABAergic transmission, such as benzodiazepines and baclofen, are commonly used to manage symptoms like spasticity and rigidity associated with cerebellar disorders.

Baclofen, a GABA-B receptor agonist, can reduce muscle spasticity and improve coordination in patients with cerebellar dysfunction. It acts by inhibiting excitatory neurotransmitter release, thereby reducing muscle hyperactivity.

Anti Cholinergic Agents

Acetylcholine is another neurotransmitter involved in motor control, and its dysregulation can contribute to movement disorders. Anti cholinergic drugs, such as trihexyphenidyl, can help alleviate symptoms like tremors and rigidity by stopping the effects of acetylcholine.

While not directly targeting cerebellar dysfunction, anti cholinergic medications can provide symptomatic relief in certain cases of movement disorders, including those affecting the cerebellum.

Dopamine Receptor Agonists

In addition to levodopa, dopamine receptor agonists like pramipexole and ropinirole are used to manage motor symptoms in cerebellar disorders. These drugs mimic the effects of dopamine in the brain and can improve coordination and reduce tremors.

Dopamine receptor agonists are often used as adjunctive therapy to levo dopa or as mono therapy in patients with mild symptoms or early-stage cerebellar disorders.

Muscle Relaxants

Muscle relaxants such as tizanidine and dantrolene may be prescribed to reduce muscle stiffness and spasticity in patients with cerebellar dysfunction. These drugs act centrally or peripherally to reduce muscle tone and promote smoother movement.

While not specific to cerebellar disorders, muscle relaxants can be beneficial in managing symptoms like muscle rigidity and spasticity, which are commonly observed in conditions affecting the cerebellum.

Conclusion

While pharmacotherapy can help alleviate symptoms and improve quality of life for individuals with cerebellar disorders like asynergia, it's essential to note that medication management should be tailored to each patient's specific needs and may require a multidisciplinary approach involving neurologists, physical therapists, and other healthcare professionals. Whatever it may be, ongoing research into novel drug targets and therapeutic approaches holds promise for the development of more effective treatments for cerebellar dysfunction in the future.

References

Ashizawa T, Xia G. Ataxia. Continuum (Minneap Minn). 2016 Aug;22 (4 Movement Disorders):1208-26. [PMC free article] [PubMed]

D'Arrigo S, Viganò L, Grazia Bruzzone M, Marzaroli M, Nikas I, Riva D, Pantaleoni C. Diagnostic approach to cerebellar disease in children. J Child Neurol. 2005 Nov;20(11):859-66. [PubMed]

Gudlavalleti A, Tenny S. StatPearls [Internet]. StatPearls Publishing; Treasure Island (FL): Oct 31, 2022. Cerebellar Neurological Signs. [PubMed]

Jimsheleishvili S, Dididze M. StatPearls [Internet]. StatPearls Publishing; Treasure Island (FL): Jul 24, 2023. Neuroanatomy, Cerebellum. [PMC free article] [PubMed]

Kular S, Cascella M. StatPearls [Internet]. StatPearls Publishing; Treasure Island (FL): Feb 5, 2022. Chiari I Malformation. [PubMed]

Mitoma H, Buffo A, Gelfo F, Guell X, Fucà E, Kakei S, Lee J, Manto M, Petrosini L, Shaikh AG, Schmahmann JD. Consensus Paper. Cerebellar Reserve: From Cerebellar Physiology to Cerebellar Disorders. Cerebellum. 2020 Feb;19(1):131-153. [PMC free article] [PubMed]

Patel S, Barkovich AJ. Analysis and classification of cerebellar malformations. AJNR Am J Neuroradiol. 2002 Aug;23(7):1074-87. [PMC free article] [PubMed]

Ramaekers VT, Heimann G, Reul J, Thron A, Jaeken J. Genetic disorders and cerebellar structural abnormalities in childhood. Brain. 1997 Oct;120 (Pt 10):1739-51. [PubMed]

Roostaei T, Nazeri A, Sahraian MA, Minagar A. The human cerebellum: a review of physiologic neuroanatomy. Neurol Clin. 2014 Nov;32(4):859-69. [PubMed]

Severino M, Huisman TAGM. Posterior Fossa Malformations. Neuroimaging Clin N Am. 2019 Aug;29(3):367-383. [PubMed]

Shen J, Shen J, Huang K, Wu Y, Pan J, Zhan R. Syringobulbia in Patients with Chiari Malformation Type I: A Systematic Review. Biomed Res Int. 2019;2019:4829102. [PMC free article] [PubMed]

Sun W, Li G, Lai Z, Lu Z, Lin Y, Peng J, Huang J, Hu K. Subacute Combined Degeneration of the Spinal Cord and Hydrocephalus Associated with Vitamin B12 Deficiency. World Neurosurg. 2019 Aug;128:277-283. [PubMed]

Thaller M, Hughes T. Inter-rater agreement of observable and elicitable neurological signs. Clin Med (Lond). 2014 Jun;14(3):264-7. [PMC free article] [PubMed]

Valente EM, Nuovo S, Doherty D. Genetics of cerebellar disorders. Handb Clin Neurol. 2018;154:267-286. [PubMed]

Verghese J, LeValley A, Hall CB, Katz MJ, Ambrose AF, Lipton RB. Epidemiology of gait disorders in community-residing older adults. J Am Geriatr Soc. 2006 Feb;54(2):255-61. [PMC free article] [PubMed]

7

Exploring Drug Therapies for Cerebellar Disorders: Focus on Asthenia

Sony Sharlet E.[1], Muralinath E.[2], Sai Hemachand N.[3] Sravani K.[4] and Guru Prasad M.[5]

[1]College of Veterinary Science, Tirupati, Andhra Pradesh, India
[2]College of Veterinary Science, Proddatur, Andhra Pradesh, India
[3]Internee, Veterinary Poly Clinic, Nellore, Andhra Pradesh, India
[4]Nuzeveedu, Andhra Pradesh, India
[5]Vaishnavi Microbial Phama Pvt. Ltd., Hyderabad, India

Abstract

The Cerebellum is referred to as the little brain and plays an important role in coordinating voluntary movements, balance and posture. Athenian is manifested by weakness and fatigue. Patients with cerebellar dysfunction frequently experience difficulties with movement coordination, resulting in an enhanced effort and energy expenditure especially during physical activities. Regarding physical therapy, rehabilitation programs focus on enhancing coordination. balance and strength through targeted exercises as well as activities. Generally no particular drugs target asthenia in s direct manner and medications nay help manage related symptoms and improve overall function. Drugs namely levo dopa or carbi dopa, Generally used in parkinsons disease, may enhance motor symptoms and decrease asthenia especially on some patients. GABAnergic modulators namely benzodiazepines and gabapentin may help alleviate symptoms of ataxia and tremors linked to cerebellar dysfunction and reduce asthenia in an indirect manner. Side effects of cerebellar disorders include dizziness, sedation and cognitive impairment which may exacerbate asthenia. Finally it is concluded that Asthenia, a common symptom of cerebellar disorders, poses a specific challenges especially on clinical management.

Introduction

The cerebellum, often referred to as the "little brain," plays a crucial role in coordinating voluntary movements, balance, and posture. Disorders impacting the cerebellum can result in a variety of debilitating symptoms, including asthenia, which is manifested by weakness and fatigue. While drug therapies for cerebellar disorders are still evolving, researchers have made significant efforts in understanding the underlying mechanisms and exploring potential treatment options.

Understanding Asthenia in Cerebellar Fisorders

Asthenia, commonly observed in cerebellar disorders, manifests as a feeling of generalized weakness, fatigue, and lack of energy. Patients with cerebellar dysfunction often experience difficulties with movement coordination, resulting in an enhanced effort and energy expenditure during physical activities. This exhaustion can significantly impair daily functioning and quality of life.

Current Treatment Approaches

Physical Therapy

One of the primary treatment modalities for cerebellar disorders, including asthenia, is physical therapy. Rehabilitation programs focus on improving coordination, balance and strength through targeted exercises as well as activities.

Occupational Therapy

Occupational therapy helps individuals with cerebellar disorders adapt to daily activities and improve functional independence. Techniques namely energy conservation strategies and ergonomic modifications can alleviate symptoms of asthenia.

Medications

Pharmacological interventions for cerebellar disorders aim to address underlying neurological mechanisms and alleviate symptoms such as asthenia. While no specific drugs target asthenia directly, medications may help manage related symptoms and improve overall function.

Drugs Acting on Cerebellar Disorders

Dopaminergic Agents

Dopamine plays an important role in motor control and coordination, making dopaminergic agents potentially beneficial for cerebellar disorders. Drugs such as levodopa/carbidopa, commonly used in Parkinson's disease, may improve motor symptoms and reduce asthenia in some patients.

GABAergic Modulators

Gamma-aminobutyric acid (GABA) is an inhibitory neurotransmitter that controls neuronal excitability. GABAergic modulators, such as benzodiazepines and gabapentin, may help alleviate symptoms of ataxia and tremors related to cerebellar dysfunction, indirectly reducing asthenia.

Cerebellar Modulators

Emerging research focuses on developing drugs that specifically target cerebellar function. Modulators of ion channels, neurotransmitter receptors, and cellular signaling pathways within the cerebellum hold promise for treating various cerebellar disorders, including asthenia.

Future Directions and Challenges

While drug therapies offer hope for managing asthenia and other symptoms of cerebellar disorders, several challenges remain:

Limited Understanding

The complex patho physiology of cerebellar disorders poses challenges in developing targeted drug therapies.

Side Effects

Many medications used for cerebellar disorders can cause adverse effects such as dizziness, sedation and cognitive impairment, which may exacerbate asthenia.

Individual Variability

Response to drug therapy varies among patients, highlighting the need for personalized treatment approaches.

Conclusion

Asthenia, a common symptom of cerebellar disorders, poses a significant challenges in clinical management. While drug therapies play a complementary role alongside rehabilitation strategies, there is a need for further research to identify novel therapeutic targets and optimize treatment outcomes. By addressing the underlying neurological mechanisms and tailoring interventions to individual patient needs, healthcare providers can better support individuals with cerebellar disorders in managing asthenia and improving overall quality of life.

References

Ashizawa T, Xia G. Ataxia. Continuum (Minneap Minn). 2016 Aug;22(4 Movement Disorders):1208-26. [PMC free article] [PubMed]

D'Arrigo S, Viganò L, Grazia Bruzzone M, Marzaroli M, Nikas I, Riva D, Pantaleoni C. Diagnostic approach to cerebellar disease in children. J Child Neurol. 2005 Nov; 20(11):859-66. [PubMed]

Gudlavalleti A, Tenny S. StatPearls [Internet]. StatPearls Publishing; Treasure Island (FL): Oct 31, 2022. Cerebellar Neurological Signs. [PubMed]

Jimsheleishvili S, Dididze M. StatPearls [Internet]. StatPearls Publishing; Treasure Island (FL): Jul 24, 2023. Neuroanatomy, Cerebellum. [PMC free article] [PubMed]

Kular S, Cascella M. StatPearls [Internet]. StatPearls Publishing; Treasure Island (FL): Feb 5, 2022. Chiari I Malformation. [PubMed]

Mitoma H, Buffo A, Gelfo F, Guell X, Fucà E, Kakei S, Lee J, Manto M, Petrosini L, Shaikh AG, Schmahmann JD. Consensus Paper. Cerebellar Reserve: From Cerebellar Physiology to Cerebellar Disorders. Cerebellum. 2020 Feb;19(1):131-153. [PMC free article] [PubMed]

Patel S, Barkovich AJ. Analysis and classification of cerebellar malformations. AJNR Am J Neuroradiol. 2002 Aug;23(7):1074-87. [PMC free article] [PubMed]

Ramaekers VT, Heimann G, Reul J, Thron A, Jaeken J. Genetic disorders and cerebellar structural abnormalities in childhood. Brain. 1997 Oct;120 (Pt 10):1739-51. [PubMed]

Roostaei T, Nazeri A, Sahraian MA, Minagar A. The human cerebellum: a review of physiologic neuroanatomy. Neurol Clin. 2014 Nov;32(4):859-69. [PubMed]

Severino M, Huisman TAGM. Posterior Fossa Malformations. Neuroimaging Clin N Am. 2019 Aug;29(3):367-383. [PubMed]

Shen J, Shen J, Huang K, Wu Y, Pan J, Zhan R. Syringobulbia in Patients with Chiari Malformation Type I: A Systematic Review. Biomed Res Int. 2019;2019:4829102. [PMC free article] [PubMed]

Sun W, Li G, Lai Z, Lu Z, Lin Y, Peng J, Huang J, Hu K. Subacute Combined Degeneration of the Spinal Cord and Hydrocephalus Associated with Vitamin B12 Deficiency. World Neurosurg. 2019 Aug;128:277-283. [PubMed]

Thaller M, Hughes T. Inter-rater agreement of observable and elicitable neurological signs. Clin Med (Lond). 2014 Jun;14(3):264-7. [PMC free article] [PubMed]

Valente EM, Nuovo S, Doherty D. Genetics of cerebellar disorders. Handb Clin Neurol. 2018;154:267-286. [PubMed]

Verghese J, LeValley A, Hall CB, Katz MJ, Ambrose AF, Lipton RB. Epidemiology of gait disorders in community-residing older adults. J Am Geriatr Soc. 2006 Feb;54(2):255-61. [PMC free article] [PubMed]

8

A Comprehensive Report on Hypothalamus Disorders Understanding Adiposogenital Syndrome

Sony Sharlet E.[1], Muralinath E.[2], Sai Hemachand N.[3] Sravani K.[4] and Guru Prasad M.[5]

[1]College of Veterinary Science, Tirupati, Andhra Pradesh, India
[2]College of Veterinary Science, Proddatur, Andhra Pradesh, India
[3]Internee, Veterinary Poly Clinic, Nellore, Andhra Pradesh, India
[4]Nuzeveedu, Andhra Pradesh, India
[5]Vaishnavi Microbial Phama Pvt. Ltd., Hyderabad, India

Abstract

Adipiso genital syndrome is otherwise known as Adiposo genital dystrophy or Frolichs syndrome. It occurs because if disorder of Hypothalamus. It is manifested by a union of obesity, under developed genitals and hormonal imbalances. It essentially influences hidden and adolescents, even though cases in adults have been reported. The causes of adipose genital syndrome include tumors, trauma and I infections The symptoms of adipose genital dystrophy are obesity, delayed or incomplete puberty, under developed or incomplete puberty. Under developed genitals and hormonal imbalances. Diagnosis is based on medical history, physical examination , laboratory tests and imaging studies namely (MRI) of the brain. Treatment is dependent upon hormone replacement therapy, diet as well as life style modifications and monitoring as well as supportive care. Finally it is concluded that Adiposo genital syndrome is a rare disease manifested by obesity. Under developed genitals and hormonal imbalances, leading to the dysfunction of the Hypothalamus.

Introduction

The hypothalamus, a small region located at the base of the brain, plays a crucial role in regulating various bodily functions, including hormone secretion, temperature regulation, sleep, and appetite. Disorders affecting the hypothalamus can have profound effects on overall health and well-being. One such disorder is Adiposogenital Syndrome, also termed as adiposogenital dystrophy or Froehlich's syndrome. This article provides an information about a comprehensive overview of Adiposogenital Syndrome, including its causes, symptoms, diagnosis, and treatment options.

Understanding the Hypothalamus

Before delving into Adiposogenital Syndrome, it's essential to understand the role of the hypothalamus. This small but mighty region behaves as a control center for many autonomic functions, serving as a link between the nervous system and the endocrine system. Among its many functions, the hypothalamus controls hunger and thirst, body temperature, sleep-wake cycles, and the release of hormones from the pituitary gland.

What is Adiposo Genital Syndrome

Adiposo genital Syndrome is a rare disorder characterized by a combination of obesity, underdeveloped genitals and hormonal imbalances. It primarily influences children and adolescents, although cases in adults have also been reported. The syndrome was first described by the German physician Carl Friedrich Otto Froehlich in 1890.

Caises

The exact cause of Adiposo genital Syndrome is not fully understood, but it is believed to result from dysfunction or damage to the hypothalamus. Potential causes include:

Tumors

Tumors in or near the hypothalamus can disrupt its normal functioning, resulting in the development of Adiposo genital Syndrome.

Trauma

Head injuries or trauma to the brain, specifically in the region of the hypothalamus, can cause damage and subsequent dysfunction.

Infections

Certain infections that affect the brain, such as encephalitis or meningitis, can damage the hypothalamus and contribute to the development of the syndrome.

Symptoms

The symptoms of Adiposo genital Syndrome can vary depending on the age of onset and severity of the condition. Common symptoms include:

Obesity

Excessive weight gain, particularly around the trunk and face, is a hallmark feature of Adiposo genital Syndrome.

Delayed PR Incomplete [UBERTY]

Children with the syndrome may experience delayed onset of puberty or incomplete sexual development.

Under Developed Genitals

In males, this may manifest as small testes and underdeveloped secondary sexual characteristics, such as facial hair growth. In females, there may be underdeveloped breasts and delayed menstruation.

Hormonal Imbalances

Adiposo genital Syndrome often involves disruptions in hormone levels, including deficiencies in growth hormone, thyroid hormones, and gonadotropins (hormones that stimulate the gonads).

Diagnosis

Diagnosing Adiposo genital Syndrome can be challenging due to its rarity and the complexity of its symptoms. A comprehensive medical evaluation, including a review of the patient's medical history, physical examination, and laboratory tests, is typically required. Imaging studies, such as magnetic resonance imaging (MRI) of the brain, may also be performed to assess the structure and function of the hypothalamus.

Treatment

Treatment of Adiposo genital Syndrome aims to address the underlying hormonal imbalances and manage associated symptoms. Depending on the specific hormone deficiencies present, treatment options may include:

Hormone Replacement Therapy

This may involve supplementation with hormones such as growth hormone, thyroid hormone, or sex hormones to correct deficiencies and promote normal development.

Diet and Lifestyle Modifications

Adopting a healthy diet and regular exercise regimen can help manage obesity and improve overall health.

Monitoring and Supportive Care

Regular monitoring of hormone levels and growth parameters is essential to track progress and adjust treatment as needed. Psychological support and counseling may also be beneficial for individuals coping with the emotional and social aspects of the syndrome.

Conclusion

Adiposo genital Syndrome is a rare disorder characterized by obesity, underdeveloped genitals and hormonal imbalances, resulting from dysfunction or damage to the hypothalamus. Although it presents significant challenges in diagnosis and management, early detection and appropriate treatment can help improve outcomes and quality of life for affected individuals. Continued research into the underlying mechanisms of the syndrome is crucial for developing more effective therapeutic approaches in the future.

References

Babinski-fröhlich syndrome. Bissonnette B, & Luginbuehl I, & Marciniak B, & Dalens B.J.(Eds.), (2006). Syndromes: Rapid Recognition and Perioperative Implications. McGraw Hill. https://accessanesthesiology.mhmedical.com/content.aspx?bookid=852§ionid=49517244

Bardet-biedlsyndrome.BissonnetteB,&LuginbuehlI,&MarciniakB,&DalensB.J.(Eds.),(2006). Syndromes: Rapid Recognition and Perioperative Implications. McGraw Hill. https://accessanesthesiology.mhmedical.com/content.aspx?bookid=852§ionid=49517256

Bardet-biedlsyndrome.BissonnetteB,&LuginbuehlI,&MarciniakB,&DalensB.J.(Eds.),(2006). Syndromes: Rapid Recognition and Perioperative Implications. McGraw Hill. https://accessanesthesiology.mhmedical.com/content.aspx?bookid=852§ionid=49517256

Bardet-biedlsyndrome.BissonnetteB,&LuginbuehlI,&MarciniakB,&DalensB.J.(Eds.),(2006). Syndromes: Rapid Recognition and Perioperative Implications. McGraw Hill. https://accessanesthesiology.mhmedical.com/content.aspx?bookid=852§ionid=49517256

Bohonowych J, Miller J, McCandless SE, Strong TV (2019). "The Global Prader-Willi Syndrome Registry: Development, Launch, and Early Demographics". Genes (Basel). 10 (9). doi:10.3390/genes10090713. PMC 6770999. PMID 31540108.

Burfeind KG, Yadav V, Marks DL. Hypothalamic Dysfunction and Multiple Sclerosis: Implications for Fatigue and Weight Dysregulation. Curr Neurol Neurosci Rep. 2016 Nov;16(11):98.

Butler MG, Manzardo AM, Forster JL (2016). "Prader-Willi Syndrome: Clinical Genetics and Diagnostic Aspects with Treatment Approaches". Curr Pediatr Rev. 12 (2): 136–66. doi: 10.2174/1573396312666151123115250. PMC 6742515 . PMID 26592417.

Cassidy SB, Schwartz S, Miller JL, Driscoll DJ (2012) Prader-Willi syndrome. Genet Med 14 (1):10-26. DOI:10.1038/gim.0b013e31822bead0 PMID: 22237428

Fermin Gutierrez MA, Mendez MD. Prader-Willi Syndrome. [Updated 2021 Aug 11]. In: StatPearls [Internet]. Treasure Island (FL): StatPearls Publishing; 2021 Jan-.

Forsythe E, Kenny J, Bacchelli C, Beales PL (2018). "Managing Bardet-Biedl Syndrome-Now and in the Future". Front Pediatr. 6: 23. doi:10.3389/fped.2018.00023. PMC 5816783. PMID 29487844.

Kanakis GA, Nieschlag E (2018). "Klinefelter syndrome: more than hypogonadism". Metabolism. 86: 135–144. doi:10.1016/j.metabol.2017.09.017. PMID 29382506.

Pacoricona Alfaro DL, Lemoine P, Ehlinger V, Molinas C, Diene G, Valette M; et al. (2019). "Causes of death in Prader-Willi syndrome: lessons from 11 years' experience of a national reference center". Orphanet J Rare Dis. 14 (1): 238. doi:10.1186/s13023-019-1214-2. PMC 6829836 . PMID 31684997.

Passone CBG, Pasqualucci PL, Franco RR, Ito SS, Mattar LBF, Koiffmann CP; et al. (2018). "Prader-Willi Syndrome: What is the General Pediatrician Supposed to do? - A Review". Rev Paul Pediatr. 36 (3): 345–352. doi:10.1590/ 1984-0462/;2018;36;3;00003. PMC 6202899. PMID 30365815.

Romito LM, Scerrati M, Contarino MF, Iacoangeli M, Bentivoglio AR, Albanese A (2003) Bilateral high frequency subthalamic stimulation in Parkinson's disease: long-term neurological follow-up. J Neurosurg Sci 47:119–128 Medline

Sanchez Jimenez JG, De Jesus O. Hypothalamic Dysfunction. [Updated 2021 Aug 30]. In: StatPearls [Internet]. Treasure Island (FL): StatPearls Publishing; 2021 Jan-

Smyth CM, Bremner WJ (1998). "Klinefelter syndrome". Arch Intern Med. 158 (12): 1309–14. doi:10.1001/archinte.158.12.1309. PMID 9645824.

Suspitsin EN, Imyanitov EN (2016). "Bardet-Biedl Syndrome". Mol Syndromol. 7 (2): 62–71. doi:10.1159/000445491. PMC 4906432. PMID 27385962.

Tuite PJ, Maxwell RE, Ikramuddin S, Kotz CM, Kotzd CM, Billington CJ; et al. (2005). "Weight and body mass index in Parkinson's disease patients after deep brain stimulation surgery". Parkinsonism Relat Disord. 11(4): 247–52. doi:10.1016/j.parkreldis.2005.01.006. PMID 15878586.